Fred Percy Briggs, Merritt Lyndon Fernald

Plant Lists

Mount Katahdin

Fred Percy Briggs, Merritt Lyndon Fernald

Plant Lists
Mount Katahdin

ISBN/EAN: 9783337417949

Printed in Europe, USA, Canada, Australia, Japan

Cover: Foto ©berggeist007 / pixelio.de

More available books at **www.hansebooks.com**

43½ Langdon St.,
Cambridge, Mass., Jan. 10, 1892,

Dear Mr. Briggs:

I was very much pleased with the Mt Ktaadn plants, and Dr. Robinson wishes me to thank you for the specimen of Lycopodium _alpinum_ for the Gray Herbarium.

I have put out the plants you requested and shall send them to you by Harriet as she returns in a few days to Orono.

I certainly hope you may have the success which should be expected on your trip next Summer. I have not yet definitely decided where I shall do my active field work in the Summer but probably somewhere in the northern part of Maine. I have several enthusiastic correspondents in various parts of the state and we are all

"Among the many interesting things I have seen this year from Maine, is one which Miss Furbish collected at Fort Fairfield twelve years ago. It proves to be _Myriophyllum alterniflorum_. (See Morong. in Torr. Bull. Aug. 1891.)

Thinking they may be of some interest to you I have made a list of the more important Ktaadn plants which I have in my records. I have given the collectors so far as I know them, but in case of common species only the earlier collectors.

I hope you will remember me again in collecting next year. I will you —

very truly yours

Merritt Lyndon Fernald.

~~Clematis verticillaris~~, DC.

    E. Br. of Penob. near Ktaadn (Scribner)

~~Nuphar Kalmianum~~, Ait.   E. Br. of Penob. (Young)

Cardamine bellidifolia, L.   Chickering, Scribner.

Arenaria Grœnlandica   all.

Geum macrophyllum, Willd.   Port Cat. 1st Ed.

Potentilla fruticosa. L.   J. A. Bailey &c. &c.

  " tridentata, Ait.   all.

Amelanchier oligocarpa.   Young. Hale.

~~Saxifraga stellaris~~, V. comosa,   Blake. Scribner.

~~Hippuris vulgaris. L.~~   E. Br. of Penob. (Goodale,

Epilobium Hornemanni, Reich.   Goodale Scribner. Bigg

Viburnum pauciflorum. Pylaie.

    Banks of the Wassataquoik (Thurber,
Eupatorium ageratoides. L.

    E. Br. of Pen. opp. "Hunts" (Thurber)

Solidago macrophylla Pursh.   Scribner.

  " virga-aurea. V. alpina   Bailey. Young &c. &c.

  " humilis. Pursh.   Blake?

~~Erigeron hyssopifolium, Mx~~
      ~~E. Br. of Penob. (Scribner)~~

Gnaphalium sylvaticum. Vill.   Scribner. Bigg.

Arnica Chamissonis, Less. Goodale. „ Briggs.

Prenanthes serpentaria V. nana. Scribner, Briggs.

„ Boottii, Gray.

Vaccinium Penn. V. angustifolium Scribner.

„ caespitosum, Mx. Thurber. &c. &c.

„ uliginosum, L. „ „ „

„ Vitis-Idaea. L. „ „ „

Arctostaphylos alpina Spr. Young. C. E. Smith, &c &c

Cassiope hypnoides, Don. „ „ „

Bryanthus taxifolius, Gray. Thurber. Blake. &c &c,

Rhododendron Lapponicum. Wahl. „ Young &c, &c.

Loiseleuria procumbens, Desv. Scribner &c, &c.

Moneses grandiflora. Salisb. Thurber, &c.

Pyrola minor L Goodale.

Diapensia Lapponica. L. Young, &c &c.

Halenia deflexa, Inseb. Young. &c.

Castilleia pallida v. septentrionalis. Briggs.

Polygonum viviparum. L. Scribner.

Betula papyrifera. V. minor „

„ glandulosa. Mx. Young &c, &c.

Salix argyrocarpa. And. Blake. Scribner.

Empetrum nigrum, L.   all.

Pinus Banksiana, Lamb.   E. Br. of Pen. (Mrs Haw

Listera cordata, R Br.   Scribner. Briggs.

" Convallarioides. Nutt.   Chickering.

Habenaria obtusata, Rich   Young.

Juncus trifidus. L.   Blake. Scribner. Brig,

Luzula spadicea. V. melanocarpa.   all.

" arcuata, Meyer.   Goodale.

" spicata, Desv.   Blake. Scribner.

Scirpus caespitosus. L.   Thurber &c &c etc.

Eriophorum alpinum. L.   ? Briggs.

Carex miliaris(?) aurea, Bailey.   Scribner (?)

" rigida. V.(?) Bigelovii, Tuck.   all.

" lenticularis, Mx.   Scribner.

" torta, Booth,   Port. Cat. 1st Ed.

" rariflora. Smith.   Goodale.

" scirpoidea. Mx.   Blake. Scribner. Briggs.

" canescens. V. alpicola.   Scribner. Briggs.

" atrata, v. ovata, Booth.   Briggs.

Hierochloe alpina. R & S.   Young. Scribner.

Phleum alpinum. L.   ? Briggs.

Deschampsia flexuosa. Trin.   all.

Poa laxa. Haenke.   Young. Scribner.

Agropyrum violaceum. Rupr.   Briggs.

~~Adiantum pedatum. L.~~   ~~Briggs.~~

Aspidium aculeatum. V. Braunei.   Goodale.

Lycopodium Selago. L.   Young. Scribner. Briggs,
„ annotinum, V. pungens,   „   ic ic, ic,
„ alpinum, L.   Briggs.

Merritt Lyndon Fernald.

Cambridge.
        Jan. 10, 1892.

DEPARTMENT OF AMERICAN ARCHAEOLOGY
WARREN K. MOOREHEAD. DIRECTOR
DOUGLAS S. BYERS. ASSISTANT DIRECTOR

October 15, 1936

Dr. M. L. Fernald
Gray Herbarium
Cambridge, Mass.

Dear Dr. Fernald:

    I have found some more lists of flora of Mt. Katahdin among the papers of Mr. Fred. P. Briggs. These may or may not be of some use. I am glad that the others were of some value. I always hesitate to throw anything away that is outside my field, as I never know whether I am throwing away something that might be of some value to other workers.

    Sincerely,

Douglas S. Byers

Douglas S. Byers

enc.
DSB/EW

Oct. 17, 1936

THE NEW ENGLAND BOTANICAL CLUB

gift 2

Phillips Andover Academy

Lichens collected on Mt. Ataado,

Cetraria Fahlunensis, Schær,
Cetraria juniperina, Ach. (on Spruce)
  "   Islandica, Ach.
  "   cuculata
Parmelia stygia, Ach
  "   centrifuga, Ach
  "   ~~tenditi~~,
Stereocaulon corallinum, Fr.
Cladonia gracilis, Fr. var. verticillata Fr.
  "   amauriocrea Schæu.
  "   squamosa,
  "   furcata, Floerk. var. subulata, Flerk
  "   rangiferina Hoff
  "   "   var alpestris, Floerk.
  "   cornucopiodes, Fr.
Umbilicaria probosidea, Dc.
  "   hyperborea, Hoff.
  "   erosa, Hoff
Lecidea   sanguinaria, Ach. (on bark of tree)
  "   contigua, Fr   (on granite.)

                Moss,
Pogonatum alpinum Roth.

KTAADN.

Orthography.

Following are the different ways I have found the name of this moun-
tain in print" Ktaadn, Ktar**d**n,, *Kataadn, Ktahdn, Kataden,* Ktaden,, Ktahd**n**,, and Katahd**i**n.   The

first and last are more common than the others, at present.   I have

taken considerable pains to find out which is the better metho**d** of

spelling, and herewith give the authority for the first, which I con-

sider the only correct way.

Hon. James Hammond Trumball who is an eminent Philo**l**ogist, and our

best authority on the Indian language, — having been lecturer on the

Indian language at Yale College, editor of "Roger Williams Key to the

Indian Language," and said to be the only man who can read Elliot's

Indian Bible, — says that th**e**s mountain is "pronounced Ktaadn by the

Indians of Maine."   This I consider decisive as the only rule for

spelling such a language is to spell it as it is pronounced by the

Aborigines.

I also quote Henry D. Thoreau in his well known book " The Maine

Woods," Dr. Charles T. Jackson, the first Geologist of Maine, in his

report "Geology of Maine": John S. Springer, in his book "Forest Life

and Forest Trees:" Capt. A. J. Farrar, in "Guide to Moosehead Lake and

the North Maine Wilderness;" Lucius L. Hubbard, in "Guide to Moose-

head Lake and Nothern Maine;" and F. E. Church, in "Camps and Tramps

about Ktaadn;" in Scribners Monthly.   I could add to this number

but this is doubtless sufficient.   Wishing to obtain as much infor-

mation as possible, I wrote to the late Joseph Nicolar, an Indian

at Old Town, who has been Govenor of that tribe, and who is the author

of a book published in 1893, entitled "The Life and Traditions of the

Red Man."   I copy the following from his letters, giving it for what

it is worth.   "The word is in the language of the Penobscots, and

ought to be spelled Ktaadn, every letter having its sound."   This

spelling accords with that giving by Trumball, for the "aa" represents

the broad sound, similar to "ar" of Nicolar's spelling.

In the vicinity of the mountain are very neatly painted guideboards

put up by the Appalachin Mountain Club, with the name spelled in this

way.

It is hoped our map-makers and our state papers may have their atten-

tion called to this matter, and hereafter spell the name of our grand

old mountain according to the original Indian pronunciation.

Derivation.

The name Ktaadn is without doubt from the Indian language. I give it
below all that I am able to find in regard to the meaning of this word.
J. Hammond Trumball says it signifies "the greatest or chief mountain and is equivalent of Kittatinny, the name of a ridge of the Alle-
ghanies." Again he says "In the Abnaki dialects the compound of
mountain names is adene. Kit or Ket means great. The Abnaki name
is Ketadene, the greatest or chief mountain."
Judge C. E. Potter, in an article on the language of the Abnakis, says
"Ktaadn is doubtless a corruption of kees, "high", and auke, "a place".
Keeraarge is a corruption of this word. The tribes eastward pronounce
their words harder and more gutterally, hence the difference in sound."

Dr. J. A. Chute, who interviewed the Delaware Indians in 1834 and
obtained his information from them, gives the meaning of Ktaadn, as
"on the high hill."
        William Willis states that his informant, Sockbasinan Indian,
said it meant "large mountain or large thing.",
Henry D. Thoreau, in his list of Indian names, without giving his
authority, states that it is said to mean "highest land."

L. L. Hubbard, who has published a list of Indian words given to lakes
streams and mountains of Maine, says it means "the biggest mountain"
from Ket or k't, "big" and the inseperable adene, "mountain."

Rev. Eugene Vetromile, missionary to the Etchemin Indians, gives it as
signifying "the greatest of mountains."

Joseph Nicolar, a Penobscot Indian, writes as follows: "The word
means highest hill. It does not mean mountain because mountain is
called watjo. It is a diffucult thing to explain to one who is not
familiar with our language, because it cuts off a great many words
which are required to express certain things. For instance: spee-gan
"high", na-ker-spee-kuk, "highest" pa-nardn-ek, "hill" na-ker-kee-
nardn-ek, "highest hill". When I was a very small boy my people
said that a hill by itself, not connected with any range of mountains,
the name of Ktardn was given to it. "

Not being familiar with the Indian language I am not able to discuss
the subject, but simply give statements as I find them. While there
is some slight difference of opinion, these writers mainly agree.

Description.

Mount Ktaadn lies about Lat. $45^o \, 53^{'N.}_1$ and Lon. 69 W.   It covers most
of Township No.3, Range 9 in Piscataquis County, Maine.   Its hight
is not far from one mile above the level of the sea.   It must be at
least thirty miles in circumference at its base, and perhaps forty, it
being very irregular.   The sides are steep and hard to climb.   The
easiest approach is from the north.   A road was cut here from the
Wissattaquoik stream to the summit of Ktaadn in 18- by Mr. F. J. Tracy
of Stacyville, and that gentleman informs me that he rode to the top
on horseback.   The deer, caribou, and moose, have followed the road
in their backward and forward tramps. and in places it has the appear-
ance of a cattle path through the forest.   The top of the mountain
consists of table lands, elevations and valleys.   One slightly
sloping plateau covers at least one fine hundred acres.   The elevations
are simply great piles of boulders the size of barrels and hogs-
heads, which look as if some giant Titan in ages past had heaped them
up one by one.   As seen at a distance the mountain shows two large
peaks.   The southern one is called Pamola in honor of an Indian Deity
or "Big Devil" which they suppose dwelt there and caused the storms
and winds.   The nothern peak is Ktaadn proper.   The pecularity about
Ktaadn that makes it different from any other mountain in the world
and adds to it so much interest, is the "basin."   This is a large
horse-shoe shaped cavity, like the crater of a volcano except that
it opens on one side toward the east.   This basin is fully two and

one half miles long by one and one half wide. One writer says there
are in it six ponds varying in size from two to ten acres. The walls
are, perhaps, two thousand feet high, nearly perpendicular, smooth,
and impossible to ascend. At little distances there are torrent
beds in which during heavy rains the water rushes down, carrying heavy
stones with it and wearing into the solid rock. One can ascend in
one of these dry beds, provided he has enough nerve, strength and
endurance. The principal danger for a party is that the one ahead
may loosen a stone which is likely to send those beneath him into the
basin and eternity. From near the center of the basin a long ridge
called the "saddle" runs to the summit, cutting the basin into.
This ridge affords an easier and less dangerous ascent than the sides.
The floor of the basin is covered with loose rocks which have fallen
from the top and sides. Hamlin gives the floor of the basin as 2900
feet above the sea. He also says that the hight of the main peak
is 2287 ft. above the basin, making it 5187 ft. above the sea level,
and Pamola is 1895 ft. above the basin. At the time I visited the
mountain the torrent beds had little streams of pure cool water
trickling down them. I went down in one of these beds and climbed
up in another. In some places I could stand erect and drink.
from the rill where the water came gurgling down vertical ledges.
I have to state here that in every place I drank water on Ktaadn it
was pure, cool, sweet and refreshing. In no place did I find any
mineral water or any that had any disagreeable qualities. We
frequently spoke of the excellence of the water, and all of the brooks

running from the mountain are po table any time of the year.

Geology.

The mountain is composed of granite. There are two varieties the red and the gray. The former appears to be at the top and the latter at the base. The country around for miles is of granite also. The first rock of a different kind that I found was melaphyr, about fifteen miles to the east. There were a number of boulders of this scattered along for a mile. I imagine it was in situ but have no proof of this. On the summit are evidences of glacial action such as worn pebbles of quartz, argyllite, etc., which shows conclusively that the ice must have passed over Ktaadn during the glacial period. There have been slides down the mountain side. Williams says that one took place on the S. W. side in 1816, 1 1/4 miles long. The east side is less than a mile. It occured between the years 1820 and . 1830. Hunters Rest is a curious place and worthy of discription. It is at the base of the mountain, on the west sideof the long ridge stretching away to the north and known as Russel mountain. It was doubtless formed by a large piece of rock breaking away from the mountain side and sliding to the foot. It is a sort of covered room open on three sides. The back side is nearly vertical, about the hight of one's head and the ceiling extends out eight or ten feet horizontally, so that the top and sides form almost a right angle like an ordinary room. It is of solid granite and looks as if it were cut out by man.

A few sticks of wood and the ashes and brands from a fire, showed that
parties had camped there.   It affords good protection from the
weather, equalling an open camp, for a hunter, but the thought of
what my chances would be if the massive roof should fall, was enough
to prevent me from especially desiring to spend a night beneath it.

Meteorology.

Ktaadn is the residence of the Indian Pamola or Big Devil, and they
have many traditions of his doings.   It is not strange that they
should have these fables as the mountain is the birthplace of storms.
One writer who has witnessed a thunder storm here says that the
lightning flashes were terriffic, and the roar of thunder reverbrating
from side to side of the basin filled one with awe.   Ktaadn is
commonly cloud capped when all around is bright and clear.   Often the
coolness on the top is sufficient to precipitate moisture, and showers
are of frequent occurrence as the condition of the fuel will testify.
I camped for three days on the north side near the timber line, and
even the dead standing trees were so wet that it was not easy to keep
a fire.   I have further evidences of showers on the mountain when
below it is clear, as I spent one cold, wet, dreary and sleepless
night there when six miles from the base of the mountain, as I learned
the next day, there was no rain.   One day in particular I remember,
clouds were forming on the mountain top, while the sun shone out of a
clear sky.   At times a vista would open for a moment and one could
catch a glimpse of some lake or stream, then it would close up again

and the world be shut out. The clouds would go scurrying past in
streaks and patches, hurried along by the wind. At one time the fog
settled down into the basin completely filling it while above it was
clear and beautiful. Standing on the brink of this cavity I gazed
down on the smoky mass impenetrable to the eye and realized for once
I was above the clouds. Thoreau had an experience so nearly like my
own that I quote his account of it.

"At length I entered within the skirts of the cloud which
seemed forever drifting over the summit, and yet would never be gone,
but was generated out of that pure air as fast as it flowed away; and
when, a quarter of a mile farther, I reached the summit of the ridge,
which those who have seen in clearer weather say is about five miles
long, and contains a thousand acres of table-land, I was deep within
the hostile ranks of the clouds, and all objects were obscured by them
Now the wind would blow me out a yard of clear sunlight, wherein I
stood; then a gray, dawning light was all it could accomplish, the
cloud line ever rising and falling with the winds intensity. Some
times it seemed as if the summit would be cleared in a few moments,
and smile in sunshine; but what was gained on one side was lost on another. It was like sitting in a chimney and waiting for the smoke
to blow away. It was, in fact a cloud factory; these were the cloud
works, and the wind turned them off done, from the cool, bare rocks."

Snow remains late in the spring and falls early in the autumn.
One party records snow and hail falling the latter part of September.
It is not easy to get a clear view of Ktaadn and many writers have
been dissapointed.   I obtained a good view of the country near
by but it was in dog-days when the air is thick, and at a distance
of thirty or forty miles everything appeared hazy gradually growing
more and more indistinct till it was lost in a blur.

Vouldst thou hear music such as ne'er wa
    planned
  For mortal ear ? Song wilder than the tun
  The Arctic utters when its waters croon
'heir angry chorus on the Norway strand,
)r where Nile thunders to a thirsty land
  With welcome sound from Mountains of th
    Moon,
  Or lone Lualaba from his lagoon
)raws down his murmurous wave ? The
    thou shouldst stand
Where dark Katahdin lifts his sea of pines
  To meet the winter storm, and lend thine ea
'o the horse ridges, where the wind entwines
  With spruce and fir, and wakes a might
    cheer,
'ill the roused forest, from its far confines,
  Utters its voice, tremendous, lone, austere.
  William Prescott Foster in the Century.

Fauna.

Deer, caribou, and moose inhabit the forests on the slopes, and deer
at least go to the very top, as I saw signs of their being all over the
mountain and even found horns which they had shed on the summit.
Bears are common around the base, and without doubt all the animals
of Northern Maine ascend to a greater or less hight. As I was
especially interested in studying the plants of this region I did not *observe*
the fauna closely so am not able to give an extended account of it.
One party noticed a weasle in the basin, near their camp.
I saw ruffed grouse as far as the timber line extended, one flock
even near the top, in the scrub. Toads were abundant in the woods
up to where the firs were but a foot high. This was our common
toad (Bufo). I saw at least two species of beetles on the summit, and
on the southern slope there were numbers of green grasshoppers
(Pevotettex glacalis, Scudd. Last but not least by any means, were
black flies. On the highest points the wind would generally blow
them away, but on the sheltered side of a rock pile they would
swarm on one. It was almost impossible to endure life unless one
was besmeared with grease containing tar, pennyoyal, or some other
odoriferous substance.

Flora.

The mountain is well wooded on the sides and at the base. A great
deal of lumbering has b%n done around it, but lumbermen say that as
soon as one begins to ascend the timber is not so good. The principal
growth is spruce, This is mixed with fir, some birch and other trees.
As one ascends, the spruce gradually gives way to fir which grows
smaller and smaller till it dwindles away to a little sprawling ever-
green that one walks over as he would boughs lying on the ground. On
the top of the mountain in the valleys is what is known as "scrub."
This consists of fir growth about four feet high and three ~~and three~~ 4
to six inches in diameter at the base. It is generally partially dead
at the top, stand close together as it can grow, and has its scraggly
branches woven together so as to form an almost impenetrable barrier.
One could scarcely penetrate a mile of it in a day. A section through
one of these trees,—if they can be called trees,—about six inches
in diameter, showed over one hundred rings so narrow that they could,
with difficulty be counted. the wood was very hard and everything
bore evidence of its slow growth. In one place on a smooth slope,
firs about four feet high were growing snug together like a hedge, and
so thickly were the branches intermingled that one of the party lay
down and rolled over and over on the top for two or three rods,
without breaking through. All vegetation hugs the ground. Willows
spread out on the rocks and run along like creepers. Nothing rises
higher than a few inches. Plants blossom with a half-inch of stem.

Sometimes the flowers are just sticking out of the ground.   In

places there are flat green patches covered with grasses, sedges, and

rushes.   The most common of these are Deschampsia caespitosa,

Carex rigida, var. bigiovia, and Juneus trifida.   The billberry

(Vaccinium uliginosum)was very abundant and one was suprised to see

such quantities of blue, bright berries in so barren a region.
(Vaccinium vitis-idea) and black crowberry
The mountain cranberry (Empetrum nigrum) were common, and these

variäies of blue, black, and red berries when gathered and stewed

with a liberal allowance of sugar made an appetizing sauce.   Other

plants which, *are* distinctly alpine and are more or less common on the

summit, are willow (Salix uva-ursa), bearberry (Arctsostaphylos

alpina),golden rod (Solidago vigaurea, var. alpina), mountain sandwort

(Arenaria groenlandica) and club moss (Lycopodium selago).   Mosses

are common under and between rocks, and black and yellow lichens, *spot* the

granite boulders.   The following list of plants grow on the mountain

All of these were observed after I had climbed some distance.   Around

the base of the mountain can be found all of the plants common

to the nothern part of the state.

Lichens.

Cetraria fahlunensis, Schaer.
Cetraria juniperina, Ach.
Cetraria cuculata,
Cetraria islandica, Ach.
Parmelia stygia, Ach.
Parmelia centrifuga, Ach.
Stereocaulon corallinum, Fr.
Cladonia gracilis, Fr. var. verticillata, Fr.
"   amaurocrea, Schaer.
"   squamosa,
"   furcata, Floerk. var. subulata, Floerk.
"   rangiferina, Hoff.
"   "   "   var. alpestris, Floerk.
"   cornucopoides, Fr.
Umbilicaria probosidea, Dc.
"   hyperborea, Hoff.
"   erosa, Hoff.
Lecidea sanguinaria, Ach.
Lecidea contigua, Fr.

Salix argyrocarpa, Anders.
Salix uva-ursi, Pursh.
Salix herbacea, L.
Empetrum nigrum, L.
Picea alba, Link.
Abies balsamea, *Miller.* .
Listera cordata, R. Brown.
Habenaria dilitata, Gray.
Habenaria obtusata, Rich.
Smilacina trifolia, Desf.
Maianthemum canadense, Desf.
Juncus filiformis, L.
Juncus, trifidus, L.
Luzula spadicea, DC. var. melanocarpa, Meyer.
Luzula arcuata, Meyer.
Luzula spicata, Desv.
Scirpus caespitosus, L.
Eriophum alpinum, L.
Carex atrata, L. var. ovata, Boott.
Carex rigida, var. biglovii, Tuck.
Carex lenticularis, Mx.
Carex torta, Boott.
Carex rariflora, Smith.
Carex scirpoidea, Mx.
Carex canescens, L. var. alpicola, Wahl.
Hierochloe alpina, R & S.
Phleum alpinum, L.
Agrostis scabra, Willd.
Agrostis canina, L. var. alpina, Oakes.
Cinna pendula, Trin.
Deschampsia flexuosa, Trin.
Poa laxa, Haenke.
Agropyrum violaceum, Lange.
Aspidium aculeatum, Swartz. var. braunii, Koch.
Osmunda regalis, L.
Lycopodium selago, L.
Lycopodium annotinum, L. var. pungens, Spreng.
Lycopodium alpinum, L.

Coptis trifolia, Salisb.
Cardamine bellidifolia, L.
Arenaria groenlandica, L.
Oxalis acetosella, L.
Nemopanthes fascicularis, Raf. .
Rhamnus alnifolia, L'Her.
Acer spicatum, Lam.
Rubus chamaemorus, L.
Geum macrophyllum, Willd.
Potentila fruticosa, L.
Potentilla tridentata, Ait.
Pyrus arbutifolia, L. f.
Pyrus americana, DC.
Amelanchier oligocarpa, Roem.
Saxifraga stellaris, L. var. comosa, Willd.
Mitella nuda, L.
Ribes prostratum, L'Her.
Epilobium hornemanni, Reich.
Circaea alpina, L.
Heracicum lanatum, Mx.
Cornus canadensis, L.
Virurnum pauciflorum, Pylaie.
Lonicera ca crulea, L.
Solidago macrophylla, Pursh.
Solidago virgaurea, L. var. alpina, Ligel.
Aster radula, Ait.
Aster acuminatus, Mx.
Gnaphalium supinum, Vill.
Arnica chamissonis, Less.
Prenanthes serpentaria, Pursh. var. nana, Gray.
Prenanthes boottii, Gray.
Campanula rotundifolia, L.
Vaccinium canadense, Kalm.
Vaccinium pennsylvanicum, Lam. var. angustifolium, Gray.
Vaccinium uliginosum, L.
Vaccinium caespitosum Mx.
Vaccinium vitis-idaea, L.
Chiogenes serpyllifolia, Salisb.
Arctostaphyllos alpina, Spreng.
Cassandra calyculata, Don.
Cassiope hypnoides, Don.
Bryanthus taxifolius, Gray.
Kalmia glauca, Ait.
Rhododendron rhodora, Don.
Rhododendron lapponicum, Wahl.
Ledum latifolium, Ait.
Loiseleuria procumbens, Lesv.
Moneses grandiflora, Salisb.
Pyrola minor, L.
Diapensia lapponica, L.
Trientalis america, Pursh.
Castilleia pallida, Kunth. var. septentrionalis, Gray.
Polygonum viviparum, L.
Betula lutea, Mx. f.
Betula papyrifolia, Ait.
Betula papyrifera, Marshall. var. minor, Tuck.
Betula glandulosa, Mx.
Alnus viridis, DC.

Routes.

There are three approaches to the mountain, one of which is usually
taken.   One of these is on the south-swest leading from the West
Branch of the Penobscot, another is from the basin on the southeast,
and the third is from the north.   Travellers choosing the first
route ascend the West Branch in a canoe with an Indian or woodsman
for a guide.   The other two roads are by train to Stacyville, then
by buck-board or foot, to the mountain about thirty miles.   For di-
rections, suggestions, and advice as to travelling, camping, etc. the
reader is reffered to Hubbard's or Farrars "Guide to Nothern Maine"
Williams says that the first party ever visiting Ktaadn ascended in
1804, going up the West Branch.

Literature.

The literature on Ktaadn is not large. There are but few really
good accounts, principally because it has not been worked up.
Jackson and Hitchcock our only two state Geologists, visited the
mountain, but their reports are meagre.   No survey has ever been
taken of it, so all distances are mere guesses.   Its height was
determined by Dr. M. C. Fernald who carried a barometer to the summit
in 1874, and found it to be 5215 feet above the sea level.   This was
found by barometre readings on the mountain compared with others in
Winn.   My own observations at Ktaadn were principally on the flora,
and the list of plants given here is all I can vouch for, although
everything found in this article is as near correct as I have been
able to ascertain.   If I have made any wrong statements it is due
to ignorance, and I should be only too glad to correct them in a
future article which I anticipate.   I shall be glad to receive information from any one regarding Ktaadn.   It is not easy to investigate
this place, as one must go a considerable distance on foot over a
rough country, carry his provisions and bedding, endure flies and
mosquitos, take the chance of stormy weather, wade brooks and streams,
climb rocks, scramble over logs, and crawl through brush.   One can
only carry food enough to last a few days, hence his stay is usually
short.   For this reason our information is meagre.
I quote from a few writers just enough to interest the reader so that
he may find the whole account and read for himself:

First a few lines from Dr. C. T. Jackson.

"Leaving our camp on the mountain side the next morning at seven o'clock, we set out for the summit of Ktaadn, travelling steadily up the side, clambering over loose boulders of granite, trap, and graywacke, which are heaped up in confusion along its course. We at *length* reached a place where it was dangerous to walk on the loose stones, and passing over the right hand side, clambered up along the dwarfish bushes that cling to the side of the mountain."

"Two of our party became discouraged on reaching this point and, there being no necessity of their accompanying us, they were allowed to return to camp. The remainder of our ascent was extremely difficult, and required no small perseverance. Our Indian guide, Louis, placed stones along the path, in order that we might more readily find the way down the mountain, and the wisdom of this precaution was fully manifested in its sequel, At ten o'clock we reashed the tableland which forms the mountain's top, and ascends gradually to the central peak. Here the wind and driving snow and hail rendered it almost impossible to proceed, but we at length reached the central peak. The true altitude of Mount Ktaadn, above the level of the sea, is a little more than one mile perpendicular elevation. It is, then evidently the highest point in the State of Maine, and is the most abrupt granite mountain in New England!"

(P 215)

Next From John S. Springer.

"Rough granite, moss-covered rocks are spread over its whole surface from the short growth upward. Blueberries and cran- berries grow far up the sides. At the time of our visit considerable snow lay on its summit and lined the walls of the great basin. The party, of course, found plenty of drink. The Avalanch brook, having ~~having~~ its source about the middle of the slide, furnished water pure as crystal. The ascent was attended with some danger and fatigue. Butwhat a view when the utmost heights are gained. What a magnificent panorama of forests, lakes, and distant mountains. The surface

variation in altitude and size all the way up to the point where it
ceases. "   (P 209)

Thoreau. gives the following account of the roughness of the mountain

"Having, slumped, scrambled, rolled, bounced, and walked by
turns, over this scraggy country, I arrived upon a side-hill, or
rather side- mountain, where rocks, gray, silent rocks, were the
flocks and herds that pastured, chewing a rocky cud at sun set.
They looked at me with hard gray eyes, without a bleat or a low.
The mountain seemed a vast aggregation of loose rocks, as if some time
it had rained rocks, and they lay as they fell on the mountain sides,
nowhere fairly at rest, but leaning on each other, all rocking stones
with cavities between, but scarcely any soil or smoother shelf.
They were the raw material of a planet dropped from an unseen quarry,
which the vast chemistry of nature would anon work up, or work down,
into the smiling and verdant plains and valleys of earth.   This was an
undone extremity of the globe; as in lignite we see coal in the pro-
cess of formation." (P 222-223)

Winthrop says in his pleasant way.-

"Ktaadn's self is finer than what Ktaadn sees. Ktaadn
is distinct, and its view is indistinct. It is a vague panorama,
a mappy, unmethodic maze of water and woods, very roomy, very vast,
very simple,— and these are capital qualities,— but also quite
monotonous. A lover of largeness and scope has the proper emotions
stirred, but a lover of variety very soon finds himself counting the .
lakes. It is a wide view, and it is a proud thing for a man six feet
or less high to feel that he himself, standing on something he himself
has climbed, and having Ktaadn under his feet for a mere convience, can
see all Maine. It does not make Maine less but the spectator more,
and that is a useful moral result. Maine's face thus exposed has
no features; there are no great mountains visible, none that seem
more than green hillocks in the distance. Besides sky, Ktaadn's
view contains only the two primeal necessities wood and water.
Nowhere have I seen such breadth of solemn forest, gloomy, were it
not for the cheerful interruption of many fair lakes and brights ways
of river linking them." (P 231)

Willaimson in his history of Maine says: "The Indians feared till
lately to visit the summit of Ktaadn.   They superstitiously supposed
it to be the summer residence of an evil spirit, called by them Pamola
who in the beginning of Snow-time rose with a great noise, and took
his flight to some unknown warmer regions.   Theyytell a story, that
seven Indians, a great many moons ago, too boldly went up the mountain
and were certainly killed by the great Pamola : for, they say, 'we
never hear of them more, and our fathers told us, an indian never
goes up to the top of Ktaadn and lives to return.'

  They say that Pamola is very great and very strong indeed;
that his head and face are like a man's, his body, shape and feet,
like an eagle's, and that he can take up a moose with one of his claws"

BIBLIOGRAPHY.

A Geographical View of the District of Maine. (1816)
                                    Joseph Whipple.
History of the State of Maine. (1832) William D. Williamson.

Reports of the Geology of the State of Maine. (1837-8-9)
                                    Charles T. Jackson.
Going to Ktaadn. Putnams Magazine. Vol.VII.p242.   Anon.

Forest Life and Forest Trees. (1851) John S. Springer.

Geology of Maine. Agriculture of Maine. Report. (1861)
                                    C. H. Hitchcock.
TheMaine Woods. (1864) Henry D. Thoreau.
The Abnaki etc
In the Open Air.   Theodore Winthrop.

Glacial Action on Mt. Ktaadn. American Journal of Science and Arts.
          Whole No. CIII. p. 27.Jan. (1872) John DeLaski.
Scientific Obsevations on Mount Ktaadn. Whig and Courier. Nov. 9,(1874)
                                    M. C. Fernald.
Camps and Tramps about Ktaadn. Scribners Monthly. Vol.XVI. No.1.May,(1878)
                                    F. E. Church.
Summer Vacations at Moosehead Lake and Vicinity. (1879)
Observation etc   Lucius L. Hubbard.
Routes to Ktaadn. Appalachia. Dec. (1881) C. E. Hamlin.
Canoe and Camera. (1882) Thomas Sedgwick Steele.

Woods and Lakes of Maine. (1884) Lucius L. Hubbard.

Guide to Moosehead Lake and Northern Maine.   Lucius L. Hubbard

Down the West Branch. (1886) Capt. A. J. Farrar.

Guide to Moosehead Lake and the North Maine Wilderness. (1889)
                              Capt. A. J. Farrar.
A Trip to Ktaadn. The Cadet. Vol.VI. No. 6. Sept.(1891)
                                    F. P. Briggs.
Mt. Ktaadn and its Flora. Botanical Gazette. Vol. XVII.No. 2. Feb. (1892)
                                    F. Lamson Scribner.
The Sacred Cow". Lewiston Journal.   L. A. Rogers.

Plants Collected on Mt. Ktaadn. Bull. Torr. Bot. Club. Vol. XIX.No. 11.Nov
                                    F. P. Briggs.

The Abnakis and their History (1866)
          Rev. Eugene Vetromile
Observations upon the Physical Geography and Geol.
of Mount Ktaadn and the Adjacent District (18
          Charles E. Hamlin

A Geographical View of the District of Maine. (1616) Joseph Whipple.

History of the State of Maine. (1832) William D. Williamson.
Reports of the Geology of Maine. (1837-8-9) Charles T. Jackson.
Going to Ktaadn. Putnams Magazine. Vol. VII. p 242.   Anon.
Forest Life and Forest Trees. (1851) John S. Springer.
Geology of Maine. Agriculture of Maine, Report. ((1861) C.H.Hitchcock
The Maine Woods.   (1864)   Henry D. Thoreau.
The Abnakis and their History. (1866) Rev. Eugene Vetromile.
In the Open Air.   Theodore Winthrop.
~~Glacial Action on Mt. Ktaadn~~.
Glacial Action on Mt. Ktaadn. American Journal of Science and Arts.
          Whole No. CIII. p. 27, Jan. 1872. John DeLaski.
Scientific Observations on Mount Ktaadn. Whig and Courier. Nov. 9, 1874
                                        M.C.Fernald.
Camps and Tramps about Ktaadn. Scribners Monthly. Vol. XVI. No 1.
                         May, 1878. F.E.Church.
Summer Vacations at Moosehead Lake and Vicinity. (1879)
                                    Lucius L. Hubbard.
Observations upon the Physical Geography and Geology of Mount
Ktaadn and the Adjacent District. (1881) C.E.Hamlin.
Routes to Ktaadn. Appalachia. Dec. (1881)   C.E.Hamlin.
Canoe and Camera. (1882) Thomas Sedgwick Steele.
Woods and Lakes of Maine. (1884)   Lucius L. Hubbard.
Guide to Moosehead Lake and _Northern_ Maine.   . L.L.Hubbard.

Down the West Branch.   (1886)   Capt. A.J.Farrar.
Guide to Moosehead Lake and the North Maine Wilderness. A.J.Farrar.
A Trip to Ktaadn. The Cadet. Vol.VI. No. 6. Sept. 1891. F.P.Briggs.
Mt. Ktaadn and its Flora. Botanical Gazette. Vol. XVII. No. 2.
                    Feb. 1892. F.Lamson Scribner.
"The Sacred Cow". Lewiston Journal.   L.A.Rogers
Plants Collected on Mt. Ktaadn. Bulletin Torrey Botanical Club
          Vol. XIX. No. 11. Nov. 1892. F.P.Briggs.

www.ingramcontent.com/pod-product-compliance
Lightning Source LLC
Chambersburg PA
CBHW022033190326
41519CB00010B/1699